1

An Introductory to Reimann Integal

Naveed Latif

Department of Mathematics,

Govt. College University, Faisalabad 38000, Pakistan

Table of Contents

Chapter 1

Introduction

"Mathematical Analysis" is generally understanding to the area of mathematics which we use several concepts of limits. In 1821, Augustin-Louis Cauchy (1789-1857) gave the following definition

"If the successive values attributed to the same variable indefinitely a fixed value, such that they finally differ from it by as little as one wishes, this latter is called the limit of all others."

The final steps in formulating the definition of limit were given by Karl Weierstrass (1815-1897).

1.1 Limits

Definition 1. Limit of function :
Let $\phi \subseteq \mathbb{R}$ and let a be cluster point of A. Function $\phi : A \rightarrow \mathbb{R}$, $l \in \mathbb{R}$ is limit of function ϕ at a, if given $\epsilon > 0$, $\exists$ a $\delta > 0$ such that if $x \in A$ and

$$| f(x) - l | < \epsilon, \quad whenever \quad | x - a | < \delta. \tag{1.1.1}$$

Sequential Criteria for limits :
This is the notion of limit in the sense of limits of sequences.

Theorem 1. (*Sequential Criteria*) *Suppose* $\varphi : S \rightarrow \mathbb{R}$ *and a is cluster point of S. Then the following propositions are equivalent:*

  (*i*) $lim_{x \rightarrow a}\varphi = l$.

  (*ii*) *For every sequence* (α_n) *in S that converges to a such that* $\alpha_n \neq a$, $\forall n \in \mathbb{N}$, *the sequence* $(\varphi(\alpha_n))$ *converges to l.*

Proof.   • $(i) \Rightarrow (ii)$:

In this part, given supposition is that $lim_{x \to a}\varphi = l$. Which means that for every $\epsilon > 0$, $\exists$ a $\delta > 0$ such that

$$|\varphi(x) - l| < \epsilon, \quad whenever \ \ 0 < |x - a| < \delta. \tag{1.1.2}$$

In part (ii), also given that $(x_n) \to a$ such that $x_n \neq a$, $\forall n \in N$, we have to prove that $\varphi(x_n) \to l$. By using (1.1.2), we take a natural number $k(\delta)$ for the given δ such that if $n \geq k(\delta)$, then $|x_n - a| < \delta$. For such x_n we must have $|\varphi(x_n) - l| < \epsilon$. Therefore,

$$|\varphi(x_n) - l| < \epsilon, \quad whenever \ \ n > k(\delta). \tag{1.1.3}$$

This shows that $\varphi(x_n)$ converges to l.

• $(ii) \Rightarrow (i)$

We prove it by using the contrapositive statement, which means that we can suppose that $\varphi(x) \not\to l$, and then we have to prove that $\varphi(x_n) \not\to l$. If $\varphi(x) \not\to l$, then $\exists$ an ε_0−neighbourhood U_{ε_0} such that there will be at least one number x_δ in $S \cap U_\delta(a)$ with $x_\delta \neq a$ such that $\varphi(x_\delta) \notin U_{\varepsilon_0}(l)$. So for every natural number n, $\frac{1}{n}$ -neighbourhood of a contains a number x_n such that

$$0 < |x_n - a| < \frac{1}{n} \quad and \ \ x_n \in S, \tag{1.1.4}$$

but

$$|\varphi(x_n) - l| \geq \varepsilon_0 \quad forall \ \ n \in \mathbb{N}. \tag{1.1.5}$$

From this, we can easily conclude that $(x_n) \to a$ in $S \ \{a\}$, but $\varphi(x_n) \not\to l$. So, we proved that if (i) is false then (ii) is also false and by using contrapositive statement of this is that if (ii) is true then (i) is true. Which is required.

$\square$

1.2   Continuity

We now start the notion of most important class of functions that is continuous functions. The term "continuous" used in the age of Newton to discuss the motion of the body or to elaborate unbroken curve, but in this time, not precisely define till to nineteenth century. After that Bernhard Bolzano in 1817 and Augustin-Louis Cauchy

1821 gave the different properties and proposed definition but still there were not be better understandable and after that Karl Weierstrass in 1870 worked on it and gave a deep and understandable idea of continuity.

Definition 2. (*Continuous function*) Suppose $S \subseteq \mathbb{R}$, also $\varphi : S \to \mathbb{R}$, and let $a \in S$. φ is continuous at a, if for every $\epsilon > 0$, $\exists\ \delta > 0$ such that

$$| \varphi(x) - \varphi(a) | < \epsilon, \quad whenever \quad | x - a | < \delta, \tag{1.2.1}$$

where $x \in S$.

If (1.2.1) is not true, then we say that φ is discontinuous at a.
A well-known functions whose are continuous are polynomial function and exponential function and so many others are too. One of discontinuous function is Dirichlet's function defined as

$$\varphi(x) = \begin{cases} 1\ , & x\ is\ rational; \\ 0, & x\ is\ irrational. \end{cases} \tag{1.2.2}$$

This function is dicontinuous at every point in $\mathbb{R}$.
Let us recall some theorems about continuity:

Theorem 2. (*Boundedness Theorem*) *Let J be closed as well as bounded interval and suppose $\varphi : J \to \mathbb{R}$ be continuous on J. Then φ is bounded on J.*

Proof. We prove it by contradiction, suppose that φ is not bounded, so by definition of unbounded, for any $n \in \mathbb{N}\ \exists$ a number $x_n \in J$ such that $|\varphi(x_n)| > n$. Since J is bounded, the sequence (x_n) is bounded. Since by Bolzano-Weierstrass Theorem says that "Every bounded sequence of real numbers has a convergent subsequence," so by using this theorem there is a subsequence $X' = (x_{n_k})$ of X such that $x_{n_k} \to x$. Sice J is closed and $x_{n_k} \in J$, so its limit is also belong to J we used here this result that if (x_n) converges and $\alpha \leq x_n \leq \beta$, $for all\ n \in \mathbb{N}$, then $\alpha \leq \lim x_n \leq \beta$. Which shows that φ is continuous at x and that's why $\varphi(x_n) \to \varphi(x)$. We know that every convergent sequence is bounded, so $\varphi(x_{n_k})$ is a convergent sequence must be bounded. But this contradicts the fact that

$$|\varphi(x_{n_k})| > n_k \geq k \quad for\ k \in \mathbb{N}. \tag{1.2.3}$$

Therefore, our supposition is wrong. Which is required. $\qquad\square$

Theorem 3. (*Maximum − Minimum Theorem*) *Suppose* $J := [\alpha, \beta]$ *be a closed and bounded interval and* $\varphi : J \to \mathbb{R}$ *be continuous on* J. *Then* φ *has an absolute maximum and an absolute minimum on* J.

Proof. Let us suppose

$$\varphi(J) := \{\varphi(x) : x \in J\}. \tag{1.2.4}$$

Since $\varphi(J)$ is bounded in $\mathbb{R}$. So $\exists$ supremum as well as infimum exist and we denote $\bar{q} := sup\varphi\{J\}$ and $\underline{q} := inf\varphi\{J\}$. We can claim that $\exists$ points q_1 and q_2 in J such that $\varphi(q_1) = \bar{q}$ and $\varphi(q_2) = \underline{q}$. We firstly establish the existence of $\bar{q}$. Since $\bar{q} = sup\varphi(J)$ if $n \in \mathbb{N}$ then $\bar{q} - \frac{1}{n}$ is not upperbound of $\varphi(J)$, so $\exists$ an element in $(x_n) \in J$ such that

$$\bar{q} - \frac{1}{n} < \varphi(x_n) \leq \bar{q} \quad forall \ n \in \mathbb{N}. \tag{1.2.5}$$

Given that J is bounded, therefore (x_n) is bounded, using Bolzano-Weierstrass Theorem $\exists$ a subsequence $(x_{n_k}) \to \bar{q}$. Because $(x_{n_k}) \in J$ and J is closed therefore $\bar{q} \in J$. Since φ is continuous at $\bar{q}$ so that $\varphi(x_{n_k}) \to \varphi(\bar{q})$. By using (1.2.5)

$$\bar{q} - \frac{1}{n_k} < \varphi(x_{n_k}) \leq \bar{q} \quad forall \ k \in \mathbb{N}, \tag{1.2.6}$$

by using Sandwich Theorem (Squeeze Theorem) $\varphi(x_{n_k}) \to \bar{q}$. Hence

$$\varphi(q_1) = lim(\varphi(x_{n_k})) = \bar{q} = sup(\varphi(J)). \tag{1.2.7}$$

Which shows that $\bar{q}$ is an absolute maximum point of φ in J.
Similarly we can prove the other part for absolute minimum. □

Theorem 4. (*Uniform Continuity Theorem*) *Suppose* $J := [\alpha, \beta]$ *be a closed and bounded interval and* $\varphi : J \to \mathbb{R}$ *be continuous on* J. *Then* φ *is also uniformly continuous on* J.

Proof. We prove it by contradiction. Suppose that φ is not uniformly continuous on J, this means that by definition of uniform continuity that $\exists \varepsilon_0 > 0$ and two sequences (u_n) and (v_n) in J such that

$$|u_n - v_n| < \frac{1}{n} \quad and \quad |\varphi(u_n) - \varphi(v_n)| \geq \varepsilon_0 \ \forall n \in \mathbb{N}. \tag{1.2.8}$$

Because J is bounded, therefore by Bolzano-Weierstrass Theorem there exists a subsequence $(u_{n_k}) \to t$ of (u_n). $t \in J$, because J is closed and using the concept that if , then $\alpha \leq limu_n \leq \beta$. It is obvious that $v_{n_k} \to t$, since

$$|v_{n_k} - t| \leq |v_{n_k} - u_{n_k}| + |u_{n_k} - t|. \tag{1.2.9}$$

$\varphi(u_{n_k}) \to \varphi(t)$ and $\varphi(v_{n_k}) \to \varphi(t)$, because that φ is continuous at point t. But this is impossible

$$|\varphi(u_n) - \varphi(v_n)| \geq \varepsilon_0 \ \forall \ n \in \mathbb{N}. \tag{1.2.10}$$

So our assumption that φ is not uniformly continuous on the closed bounded interval J then φ is not continuous at point $t \in J$. Which is required. □

Lipschitz Functions :

The above theorem is valid if J is closed as well as bounded but if interval is not closed and bounded, then sometime it is difficult to know that the function is uniformly continuous or not. Then Lipschitz functions give a sufficient condition for uniform continuity.

Definition 3. Suppose $S \subseteq \mathbb{R}$ and $\varphi : S \to \mathbb{R}$. If $\exists$ a constant $L > 0$ such that

$$|\varphi(u) - \varphi(v)| \leq L|u - v| \ \forall \ u, v \in S, \tag{1.2.11}$$

then φ is said to be Lipschitz function on S, it is also well-known as Lipschitz condition.

The geometrical behaviour of Lipschitz function is that the slope of the line segments joining the points $(u, \varphi(u))$ and $(v, \varphi(v))$ over J are bounded by L, i.e.,

$$|\frac{\varphi(u) - \varphi(v)}{u - v}| \leq L, \quad u, v \in J, u \neq v. \tag{1.2.12}$$

Theorem 5. *(Lipschitz $-$ Uniform Continuous) If $\varphi : S \to \mathbb{R}$ is a Lipschitz function, then φ is uniformly continuous on S.*

Proof. By using (1.2.11) because φ is Lipschitz function, then for given $\epsilon > 0$, we set $\delta := \epsilon L$. If $u, v \in S$ satisfying $|u - v| < \delta$, then

$$|\varphi(u) - \varphi(v)| < L.\frac{\epsilon}{L} = \epsilon, \tag{1.2.13}$$

which shows the uniform continuity. □

1.3 Differentiability

When the derivative is introduced, it is easily to see that limit of the difference of this quotient is equal to the slope of a tangent line or we can say that the axis parallel to the x-axis is elaborate with time and the axis parallel to the y-axis is elaborated with distance which is in fact the instantaneous velocity. But the integral is not so easily interpreted: Why is the area under the curve in any way related to the antiderivative? It really was a work of genius by Newton and Leibnitz to see that connection.

Definition 4. (*Derivative*) Suppose $J \subseteq \mathbb{R}$ and $\varphi : J \to \mathbb{R}$, and let $a \in J$. φ is differentiable at a, if for given $\epsilon > 0$, $\exists \ \delta(\epsilon) > 0$ such that if $x \in J$ satisfying $0 <| x - a |< \delta(\epsilon)$, then

$$|\frac{\varphi(x) - \varphi(a)}{x - a} - l| < \epsilon. \tag{1.3.1}$$

Theorem 6. (*Derivative − Continuous*) If $\varphi : J \to \mathbb{R}$ has derivative at $a \in J$, then φ is continuous at a.

Proof. Take

$$\varphi(u) - \varphi(a) = \left(\frac{\varphi(u) - \varphi(c)}{u - a} \right)(u - a), \tag{1.3.2}$$

$\forall u \in J$ and also $u \neq a$. Since φ' exists and using product rule of limit to get that

$$lim_{u \to a} (\varphi(u) - \varphi(a)) \tag{1.3.3}$$
$$= lim_{u \to a} \left(\frac{\varphi(u) - \varphi(c)}{u - a} \right) lim_{u \to a} (u - a) \tag{1.3.4}$$
$$= \varphi'(a).0 = 0. \tag{1.3.5}$$

This means that

$$lim_{u \to a} \varphi(u) = \varphi(a), \tag{1.3.6}$$

which shows the continuity at point a. □

By taking algebraic combinations of functions of the form $x \to |x - a|$, it is not difficult to construct continuous functions that do not have a derivative at a finite or even in countable points. In 1872, Karl Weierstrass gave an example that a function is continuous everywhere but it is not differentiable at any point.

Mean Value Theorem:

The Mean Value Theorem gives a relation between values of the function and the values of its derivative. which is one of the most important result in mathematical analysis.

Theorem 7. (*Interior Extremum Theorem*) *Let a be a point of $J°$ at which $\varphi :$ $J \to \mathbb{R}$ has a relative extremum. If $f'(a)$ exists, then $f'(c) = 0$.*

Proof. We prove this theorem for relative extremum, this means for relative maximum and relative minimum. First we prove it for relative maximum at a.

- If $\varphi'(a) > 0$, then $\exists$ a neighbourhood $U \subseteq J$ of a such that

$$\frac{\varphi(u) - \varphi(a)}{u - a} \quad for \quad u \in U, u \neq a. \tag{1.3.7}$$

- If $u \in U$ and $u > a$, so we can get

$$\varphi(u) - \varphi(a) = (u - a) . \frac{\varphi(u) - \varphi(a)}{u - a} > 0. \tag{1.3.8}$$

But this leads to contradiction the fact that φ has a relative maximum at a. So for this reason, we can not take $\varphi'(a) > 0$.
Similarly, we can prove that $\varphi'(a) < 0$.
By using these both facts that $\varphi'(a) \not> 0$ and $\varphi'(a) \not< 0$, combining these results, we must get that $\varphi'(a) = 0$.
Similarly, we can prove for the relative minimum. □

From this theorem, we conclude that $\varphi : J \to \mathbb{R}$ is continuous on J and φ has a relative extremum at $a \in J°$. Then f' does not exist or equal to 0.
We note that if $\varphi(x) := |x|$ on $J := [-1, 1]$, then φ has an interior minimum at $x = 0$, but f' does not exit at $x = 0$.

Theorem 8. (*Rolle's Theorem*) *Suppose that φ is continuous on a closed interval $J := [\alpha, \beta]$, that*

- *f' exists at every point on $J°$,*

- *$\varphi(\alpha) = \varphi(\beta) = 0$.*

Then $\exists$ at least one point $a \in (\alpha, \beta)$ such that $\varphi'(a) = 0$.

Proof. There will be two cases arise:

- If φ vanishes at all points on J.

- If φ does not vanish at all points on J.

In the first case, this is obviously true if φ vanishes on all points in J.

In second part, we suppose that φ does not vanish at all points in J. This function φ have may be positive values or negative values, so if it has negative values then we can take $-\varphi$ for positive value. So, we can take φ assume to be positive values. We know that if φ is continuous on any closed and bounded interval then φ has maximum value and minimum value at some point $a \in J$, so the function φ attains the value $sup\{\varphi(u) : u \in J\}$ at some point in J. Also given data that $\varphi(\alpha) = \varphi(\beta) - 0$, this logical reason shows that the point $a \in (\alpha, \beta)$, implies that $\varphi'(a)$ exists.

By using Interior-Extremum Theorem, we can conclude that for φ has a relative maximum value at a, then $\varphi'(a) = 0$. $\qquad\qquad\qquad\qquad\qquad\qquad\qquad\qquad\qquad$ $\square$

The geometrical behaviour of the Mean Value Theorem shows that tangent line at some point is parallel to the line segment through the point $(\alpha, \varphi(\alpha))$ and $(\beta, \varphi(\beta))$.

Theorem 9. (*Mean Value Theorem*) *Suppose φ is continuous on closed interval* $J := [\alpha, \beta]$,

- *φ has a derivative in J°,*

- *then $\exists$ at least one point $a \in (\alpha, \beta)$*

such that

$$\frac{\varphi(\beta) - \varphi(\alpha)}{\beta - \alpha} = \varphi'(a). \qquad\qquad (1.3.9)$$

Proof. Let us define a function

$$\Lambda(u) := \varphi(u) - \varphi(\alpha) - \frac{\varphi(\beta) - \varphi(\alpha)}{\beta - \alpha}(u - \alpha). \qquad\qquad (1.3.10)$$

The function Λ can be described as the difference of φ and the function whose graph is line segment joining the points $(\alpha, \varphi(\alpha))$ and $(\beta, \varphi(\beta))$.

Now all assumptions of Rolle's Theorem is true i.e.,

- φ is continuous on J.

- φ is differentiable on J^0.

- $\varphi(\alpha) = \varphi(\beta) = 0$.

Therefore, $\exists$ a point $a \in (\alpha, \beta)$ such that

$$\Lambda'(a) = \varphi'(a) - \frac{\varphi(\beta) - \varphi(\alpha)}{\beta - \alpha}. \tag{1.3.11}$$

Some rearrangements, we can get

$$\varphi(\beta) - \varphi(\alpha) = \varphi'(a)\,(\beta - \alpha). \tag{1.3.12}$$

$\square$

The Mean Value Theorem allows to draw this conclusion about the nature of function φ from information about its derivative.

Theorem 10. (*Application of Mean Value Theorem*) *Suppose that*

- φ *is continuous on closed interval* $J := [\alpha, \beta]$.

- φ *is differentiable on* J°,

- $\varphi'(x) = 0$, *for* $x \in (\alpha, \beta)$.

Then φ *is constant on* J.

Proof. We want to prove that $\varphi(u) = \varphi(\xi)$ for every $u \in J$. We can make an interval $J' := [\xi, u]$ if $\xi < u$. Now we apply the well-known Mean Value Theorem on the interval J' we get

$$\frac{\varphi(u) - \varphi(\xi)}{u - \xi} = \varphi'(a), \quad for\,some\,point\ a \in (\xi, u). \tag{1.3.13}$$

Given data that $\varphi'(a) = 0$ which imply that $\varphi(u) - \varphi(\xi) = 0$.
Hence, $\varphi(u) = \varphi(\xi)$ for any $u \in J$. $\square$

Theorem 11. (*Cauchy Mean Value Theorem*) *Suppose* ϕ *and* φ *be functions on* $[\alpha, \beta]$ *such that*

- ϕ *and* φ *be continuous functions on* $[\alpha, \beta]$

- ϕ *and* φ *be functions on* (α, β), *and assume that* $\varphi' \neq 0$, *for all* $x \in (\alpha, \beta)$.

Then $\exists\, a \in (\alpha, \beta)$ *such that*

$$\frac{\phi(\beta) - \phi(\alpha)}{\varphi(\beta) - \varphi(\alpha)} = \frac{\phi'(a)}{\varphi'(a)}. \tag{1.3.14}$$

Proof. Let us define a function

$$\Omega(u) := \frac{\phi(\beta) - \phi(\alpha)}{\varphi(\beta) - \varphi(\alpha)} \left(\varphi(u) - \varphi(\alpha)\right) - \left(\phi(u) - \phi(\alpha)\right) \quad for \ u \in [\alpha, \beta], \quad (1.3.15)$$

this function can be defined if $\varphi(\alpha) \neq \varphi(\beta)$, and this is true because all the conditions of Rolle's Theorem is true then by using Rolle's Theorem, if $\varphi'(u) \neq 0$, for all $u \in (\alpha, \beta)$, then $\varphi(\alpha) \neq \varphi(\beta)$.

- Ω is continuous on $[\alpha, \beta]$.

- Ω is differentiable on (α, β).

- $\Omega(\alpha) = \Omega(\beta) = 0$.

So by using well-known Rolle's Theorem that $\exists$ a point $a \in (\alpha, \beta)$ such that

$$\frac{\phi(\beta) - \phi(\alpha)}{\varphi(\beta) - \varphi(\alpha)} \varphi'(a) - \phi'(a) = \Omega'(a) = 0. \quad (1.3.16)$$

Now, dividing by $\varphi'(a) \neq 0$ we get the required result. $\qquad \square$

Chapter 2

Reimann Integral

2.1 Reimann Sum

In 1850, a beautiful turn in mathematical analysis is the work by Cauchy and after that soon the work of Bernhard Reimann. The idea of integration in starting one was completely divorced from the derivative and instead use the notion "area under the curve" as a starting point for constructing a rigorous definition of the integral.

The Reimann integral is today a very important notion in real analysis and also in introductory to calculus. This is used in this way that the function φ on $[\alpha, \beta]$, we divide this interval into small subintervals. By using each subinterval $[u_{l-1}, u_l]$, we choose any value $a_l \in [u_{l-1}, u_l]$ and then we find the value of $\varphi(a_l)$. Geometrical behavior of this is that a row of thin rectangles formed the area between φ and the horizontal axis. The area of such rectangle is defined as $\varphi(a_l)(u_l - u_{l-1})$ and Reimann sum is the total area of all such rectangles

$$Reimann\ Sum := \sum_{l=1}^{n} \varphi(a_l)(u_l - u_{l-1}). \tag{2.1.1}$$

In this notion with understanding that area below the horizontal axis is assigned to be -ve.

Reimann Sum can be accurated with increasing the number of rectangles with less area means that we divide subintervals in a very small that tends to zero. We use this in limiting process and if this limit exists then say that is Reimann's definition of $\int_{\alpha}^{\beta}$.

We introduced before the notion of derivative, it is not so tough to see that the limit of the difference quotient should be equal to to the slope of tangent line or when the axis parallel to x-axis is time and the axis parallel to y-axis is distance which is

11

the instantaneous velocity. In fact, integration, in the way of area or volume, was discovered first, many centuries before, Archimedes was involved in trying to compute the volume of a wine cask.

Definition 5. Suppose φ is bounded in a closed and bounded interval $[\alpha, \beta]$.

- P is said to be a partition of the interval $[\alpha, \beta]$ into subintervals and this set P is a finite set. If we take another partition Q such that $P \subseteq Q$, such partition is called refinement of P.

- We define partition P as $P = \{u_0, u_1, u_2, ..., u_n\}$ where $\alpha = u_0 < u_1 < u_2 < ... < u_n = \beta$. Then

$$U(\varphi, P) = \sum_{l=1}^{n} sup\{\varphi(u) : u_{l-1} \leq u \leq u_l\} (u_l - u_{l-1}) \qquad (2.1.2)$$

$$L(\varphi, P) = \sum_{l=1}^{n} inf\{\varphi(u) : u_{l-1} \leq u \leq u_l\} (u_l - u_{l-1}) \qquad (2.1.3)$$

are the upper and lower Reimann sums with respect to φ and P.

In particular partition P , it is clear that $U(\varphi, P) \geq L(\varphi, P)$.

Lemma 1. *If $P \subseteq Q$, where Q be a refinement of Q then*

$$L(\varphi, P) \leq L(\varphi, Q) \quad and \quad U(\varphi, P) \leq U(\varphi, Q). \qquad (2.1.4)$$

Proof. Since Q be a refinement of P means that subintervals in partition Q is larger in number than subintervals in partition P. So, we prove this lemma for a case if one point t is adding in the subinterval $[u_l, u_{l-1}]$ partition of P.
Then the lower sum is

$$\varphi(u_l)(u_l - u_{l-1}) \qquad (2.1.5)$$
$$= \varphi(u_l)(u_l - t) + \varphi(u_l)(t - u_{l-1})$$
$$\leq \gamma_l'(u_l - t) + \gamma_l''(t - u_{l-1}), \qquad (2.1.6)$$

where

$$\gamma_l' = inf\{\varphi(u) : u \in [t, u_l]\} \quad and \quad \gamma_l'' = inf\{\varphi(u) : u \in [u_{l-1}, t]\} \qquad (2.1.7)$$

are each necessarily as greater or greater than $\varphi(u_l)$.
By using induction method, we can prove that $L(\varphi, P) \leq L(\varphi, Q)$.
Similarly we can prove it for upper sum as well. $\qquad\qquad\square$

Lemma 2. *If R_1 and R_2 be two any partitions of $[\alpha, \beta]$, then*

$$L(\varphi, R_1) \leq U(\varphi, R_2). \tag{2.1.8}$$

Proof. If we introduce the common refinement $R = R_1 \cup R_2$, then clearly $R \subseteq R_1$ and and $R \subseteq R_2$. Then

$$L(\varphi, R_1) \leq L(\varphi, R) \leq U(\varphi, R) \leq U(\varphi, R_2). \tag{2.1.9}$$

$\square$

2.2 Reimann Integral

We can estimate upper sum as the overestimate and the lower sum is underestimate and if these both upper and lower sums are very nearly equal or you can say approaches to be equal or their differences tends to zero then we say that any function φ is integrable.

Definition 6. (*Upper and Lower Integrals*)
Suppose $\varphi : [\alpha, \beta] \to \mathbb{R}$ be a bounded function. We represent Reimann upper integral as

$$\overline{\int_\alpha^\beta} := inf\{U(\varphi, P) : P \, is \, a \, partition \, of \, [\alpha, \beta]\}. \tag{2.2.1}$$

We represent Reimann lower integral as

$$\underline{\int_\alpha^\beta} := sup\{L(\varphi, P) : P \, is \, a \, partition \, of \, [\alpha, \beta]\}. \tag{2.2.2}$$

Definition 7. (*Reimann Integral*)
Suppose $\varphi : [\alpha, \beta] \to \mathbb{R}$ be a bounded function. If

$$\int_\alpha^\beta \varphi := \overline{\int_\alpha^\beta} \varphi = \underline{\int_\alpha^\beta} \varphi, \tag{2.2.3}$$

then we say that φ is integrable on $[\alpha, \beta]$.

We can define Reimann integral for unbounded function φ but be careful for unbounded function.

Theorem 12. (*Properties of Reimann Integrals*)
Suppose $\phi, \varphi : [\alpha, \beta] \to \mathbb{R}$ be Reimann integrable functions on $[\alpha, \beta]$. Then

1. *Function $\phi + \varphi$ be Reimann integrable on $[\alpha, \beta]$ and*

$$\int_\alpha^\beta (\phi + \varphi) := \int_\alpha^\beta \phi + \int_\alpha^\beta \varphi. \qquad (2.2.4)$$

2. *If $\eta \in \mathbb{R}$ then $\eta\phi$ be Reimann integrable on $[\alpha, \beta]$ and*

$$\int_\alpha^\beta (\eta\phi) := \eta \int_\alpha^\beta \phi. \qquad (2.2.5)$$

3. *Function $\phi - \varphi$ be Reimann integrable on $[\alpha, \beta]$ and*

$$\int_\alpha^\beta (\phi - \varphi) := \int_\alpha^\beta \phi - \int_\alpha^\beta \varphi. \qquad (2.2.6)$$

4. *For $\phi(u) \geq 0$, for all $u \in [\alpha, \beta]$, then*

$$\int_\alpha^\beta \phi \geq 0. \qquad (2.2.7)$$

5. *For $\phi(u) \leq \varphi(u)$, for all $u \in [\alpha, \beta]$, then*

$$\int_\alpha^\beta \phi \leq \int_\alpha^\beta \varphi. \qquad (2.2.8)$$

6. *If $\exists \eta \in \mathbb{R}$ such that $\phi(u) = \eta$ for all $u \in [\alpha, \beta]$, then*

$$\int_\alpha^\beta \phi = \eta(\beta - \alpha). \qquad (2.2.9)$$

7. *Suppose $\gamma, \delta \in \mathbb{R}$ such that $\gamma \leq \alpha \leq \beta \leq \delta$, then $[\gamma, \delta]$ contains $[\alpha, \beta]$. Let us define a function $\Psi(u) := \phi(u)$ for all $u \in [\alpha, \beta]$ and $\Psi(u) := 0$ otherwise. Then ϕ be Reimann integrable on $[\gamma, \delta]$, and*

$$\int_\gamma^\delta \Psi = \int_\alpha^\beta \phi. \qquad (2.2.10)$$

8. *Let $\gamma \in \mathbb{R}$ such that $\alpha < \gamma < \delta$. Then $\phi\mid_{[\alpha,\gamma]}$ and $\phi\mid_{[\gamma,\beta]}$ be Reimann integrable on $[\alpha,\gamma]$ and $[\gamma,\beta]$ respectively and also*

$$\int_\alpha^\beta \Psi = \int_\alpha^\gamma \phi\mid_{[\alpha,\gamma]} + \int_\gamma^\beta \phi\mid_{[\gamma,\beta]}. \tag{2.2.11}$$

Proof. 1. We introduced here a tagged partition in this manner is that we take a functional value inside every subinterval which is in fact a partition of the interval. Let $\dot{P}$ be tagged partition of $[\alpha,\beta]$ such that $\dot{P} = \{(u_{l-1}, u_l, t_l)\}_{l=1}^n$ then we can easily seen that

$$S\left(\phi + \varphi; \dot{P}\right) = S\left(\phi; \dot{P}\right) + S\left(\varphi; \dot{P}\right), \tag{2.2.12}$$

where S be a Reimann Sum for tagged partition. We also use this definition of Reimann integral in the sense of ϵ and δ that "any function $\phi : [\alpha,\beta]$ be Reimann integrable on $[\alpha,\beta]$ if $\exists$ a number l such that for every $\epsilon > 0$, $\exists \delta_\epsilon > 0$ such that if $\dot{P}$ be a tagged partition with $\parallel \dot{P} \parallel < \delta_\epsilon$, then"

$$|S\left(\phi; \dot{P}\right) - l| < \epsilon. \tag{2.2.13}$$

By using this definition, since ϕ and φ be Reimann integrable functions on $[\alpha,\beta]$ then

$$|S\left(\phi; \dot{P}\right) - \int_\alpha^\beta \phi| < \frac{\epsilon}{2} \quad and \quad |S\left(\varphi; \dot{P}\right) - \int_\alpha^\beta \varphi| < \frac{\epsilon}{2}. \tag{2.2.14}$$

Consider

$$|S\left(\phi + \varphi; \dot{P}\right) - \left(\int_\alpha^\beta \phi + \int_\alpha^\beta \varphi\right)|$$
$$= |S\left(\phi; \dot{P}\right) + S\left(\varphi; \dot{P}\right) - \int_\alpha^\beta \phi - \int_\alpha^\beta \varphi|$$
$$\leq |S\left(\phi; \dot{P}\right) - \int_\alpha^\beta \phi| + |S\left(\varphi; \dot{P}\right) - \int_\alpha^\beta \varphi|$$
$$< \frac{\epsilon}{2} + \frac{\epsilon}{2} = \epsilon. \tag{2.2.15}$$

From this we conclude the required result.
2. Since $\dot{P}$ be a tagged partition of $[\alpha,\beta]$, then by using the concept of series we can easily get

$$S(\eta\,\phi; \dot{P}) := \eta\, S(\phi; \dot{P}). \tag{2.2.16}$$

By using the definition of Reimann integral in the sense of $\epsilon - \delta$ and tagged partition that for every $\epsilon > 0$, there exists a number $\delta_\epsilon > 0$ such that $\| \dot{P} \| < \delta_\epsilon$ then

$$|S(\phi; \dot{P}) - \int_\alpha^\beta \phi| < \frac{\epsilon}{\eta}. \tag{2.2.17}$$

Now take

$$|S(\eta\phi; \dot{P}) - \int_\alpha^\beta \eta\phi|$$
$$= \eta|S(\phi; \dot{P}) - \int_\alpha^\beta \phi| < \eta\frac{\epsilon}{\eta} = \epsilon, \tag{2.2.18}$$

since $\epsilon > 0$ is taken to be arbitrary so $\eta\phi$ is Reimann integral.

5. We prove it by using triangular inequality on (2.2.14) we have

$$\int_\alpha^\beta \phi - \frac{\epsilon}{2} < S(\phi; \dot{P}) \quad also \quad S(\varphi; \dot{P}) < \int_\alpha^\beta \phi + \frac{\epsilon}{2}. \tag{2.2.19}$$

By using the notion of series or you can say sum that

$$S(\phi; \dot{P}) \le S(\varphi; \dot{P}) \quad whenever \quad \phi \le \varphi, \tag{2.2.20}$$

therefore

$$\int_\alpha^\beta \phi \le \int_\alpha^\beta \varphi + \epsilon. \tag{2.2.21}$$

Since $\epsilon \to 0$, so the required result obtained. $\qquad\square$

Theorem 13. (*Bounded Theorem*) $\phi : [\alpha, \beta] \to \mathbb{R}$ *bounded function is an integrable function* $\Leftrightarrow$ *there exists a partition R such that*

$$U(\phi; R) - L(\phi' R) < \epsilon. \tag{2.2.22}$$

Proof. • ($\Leftarrow$) We prove it by contradiction to assume that $U(\phi; R) \ne L(\phi; R)$. So

$$U(\phi; R) - L(\phi; R) = \epsilon > 0. \tag{2.2.23}$$

By supposition there exists R such that

$$U(\phi; R) - L(\phi; R) < \epsilon. \tag{2.2.24}$$

We know that

$$L(\phi; R) \le L(\phi) \quad and \quad U(\phi) \le U(\phi; R) \tag{2.2.25}$$

here $U(\phi)$ be upper integral and $L(\phi)$ be lower integral then

$$U(\phi) - L(\phi) \leq U(\phi; R) - L(\phi; R) < \epsilon = U(\phi) - L(\phi), \qquad (2.2.26)$$

which contradicts our supposition.

- $(\Rightarrow)$ We can pick N, M be two partitions and suppose for given $\epsilon > 0$,

$$U(\phi; N) \leq L(\phi; M \cup N) \leq U(\phi; M \cup N) \leq U(\phi; M), \qquad (2.2.27)$$

hence

$$
\begin{aligned}
U(\phi; M \cup N) &- L(\phi; M \cup N) \\
&\leq U(\phi; M) - L(\phi; N) \\
&= \left(U(\phi; M) - \int_\alpha^\beta \phi \right) + \left(\int_\alpha^\beta \phi - L(\phi; N) \right) < \epsilon, \qquad (2.2.28)
\end{aligned}
$$

which is required.

$\square$

The next theorem is the relation between continuity and integrability. We know that all continuous functions are not differentiable and it is difficult to check the differentiability condition than for it to be integrable, So, it is easy for us to see that this function is integrable or not to just check the continuity about that function. Next theorem is very powerful theorem to check any function is integrable or not :

Theorem 14. *(Continuity − Integrability)*
If any function $\lambda : [\alpha, \beta] \to \mathbb{R}$ be continuous then it is integrable.

Proof. Since we know that if λ be continuous function on closed as well as bounded interval then λ be uniform continuous function on $[\alpha, \beta]$. So definition of continuity, for every $\epsilon > 0$, $\exists \delta > 0$ such that

$$|\lambda(u_l) - \lambda(u_m)| < \frac{\epsilon}{\beta - \alpha} \quad whenever \quad |u_l - u_m| < \delta. \qquad (2.2.29)$$

We can choose such partition R on which the length of all subintervals $[u_{l-1}, u_l]$ less than δ.
Because λ is continuous on $[\alpha, \beta]$ so sup and inf of λ exists on each $[u_{l-1}, u_l]$ which are y_l and z_l respectively, then

$$|y_l - z_l| \leq |u_l - u_{l-1}| < \delta, \qquad (2.2.30)$$

so

$$|\lambda(y_i) - \lambda(z_l)| < \frac{\epsilon}{\beta - \alpha}, \tag{2.2.31}$$

hence

$$U(\lambda; R) - L(\lambda; R) = \sum_{l=1}^{n} (\lambda(y_l) - \lambda(z_l))(u_l - u_{l-1})$$

$$< \sum_{l=1}^{n} \frac{\epsilon}{\beta - \alpha}(u_l - u_{l-1}) = \frac{\epsilon}{\beta - \alpha}(u_n - u_0) = \epsilon \tag{2.2.32}$$

the second last inequality is the "telescoping sum", i.e., all the terms will be cancel just first and last are remaining. □

But there are many discontinuous functions which are integrable like "Thomae function" is discontinuous but integrable but "Dirichlet function" is discontinuous as well as not integrable on $[0, 1]$.

Example 1. Thomae function is defined as:

$$\psi(u) = \begin{cases} 0, & u \in \mathbb{Q}'; \\ \frac{1}{q}, & u \in \mathbb{Q}, \end{cases} \tag{2.2.33}$$

here $\mathbb{Q}'$ is irrational set and $u \in \mathbb{Q}$ means that $u = \frac{p}{q}$ in lowest form.
This function is continuous on every irrational numbers but in the whole, Thomae function is discontinuous on $\mathbb{R}$. We can check that this is integrable function.

In fact, bounded function on a closed and bounded interval is integrable iff the set of discontinuities have measure zero, but it goes to Lebegue integration which is rather different approach and different wide concepts, I will not be to do in this monograph on that direction.
We now establish the Cauchy Criteria about integrability which will then help to prove for Squeeze (Sandwhich) Theorem and then by using Squeeze Theorem we obtain the Reimann integrability of several classes of functions like Step, Continuous and also monotone functions.

Theorem 15. *(Cauchy Criterion)*
The function ζ be Reimann integrable on $[\alpha, \beta] \Leftrightarrow$ for every $\epsilon > 0, \exists v_\epsilon$ such that $\dot{G}$ and $\dot{H}$ be any tagged partitions of $[\alpha, \beta]$ with $\| \dot{G} \| < v_\epsilon \| \dot{H} \| < v_\epsilon$, then

$$|S(\zeta; \dot{G}) - S(\zeta; \dot{H})| < \epsilon. \tag{2.2.34}$$

Proof. • ($\Rightarrow$) Suppose ζ is Reimann integrable with its limit l, suppose $\vartheta_\epsilon := \frac{\delta_\epsilon}{2} > 0$ be such that if $\dot{G}$ and $\dot{H}$ are tagged partitions such that $\| \dot{G} \| < v_\epsilon \| \dot{H} \| < v_\epsilon$, then

$$|S(\zeta; \dot{G}) - l| < \frac{\epsilon}{2} \quad and \quad |S(\zeta; \dot{H}) - l| < \frac{\epsilon}{2}. \qquad (2.2.35)$$

Then

$$|S(\zeta; \dot{G}) - S(\zeta; \dot{H})| \leq |S(\zeta; \dot{G}) - l + l - S(\zeta; \dot{H})|$$
$$\leq |S(\zeta; \dot{G}) - l| + |l - S(\zeta; \dot{H})| < \tfrac{\epsilon}{2} + \tfrac{\epsilon}{2} = \epsilon. \qquad (2.2.36)$$

• ($\Leftarrow$) For every $n \in \mathbb{N}$, suppose $\delta_n > 0$ such that $\| \dot{G} \| < v_\epsilon \| \dot{H} \| < v_\epsilon$, then

$$|S(\zeta; \dot{G}) - S(\zeta; \dot{H})| < \frac{1}{n}. \qquad (2.2.37)$$

We can assume that $\delta_n \geq \delta_{n+1}$ for $n \in \mathbb{N}$.
For every $n \in \mathbb{N}$, suppose $\dot{G}_n$ be tagged partition with $\| \dot{G}_n \| < \delta_n$. Therefore, if $m > n$ then $\| \dot{G}_m \| < \delta_n$ and $\| \dot{G}_n \| < \delta_n$, so

$$|S(\zeta; \dot{G}_n) - S(\zeta; \dot{G}_m)| < \frac{1}{n} \quad for \ m > n. \qquad (2.2.38)$$

Which shows that $\left(S(\zeta; \dot{G}_m) \right)_{m=1}^{\infty}$ is Cauchy sequence in $\mathbb{R}$. We know that every Cauchy sequence in $\mathbb{R}$ converges and suppose that $B := lim \left(S(\zeta; \dot{G}_m) \right)$. If we take limit $m \to \infty$ to (2.2.38), we get

$$|S(\zeta; \dot{G}_m) - B| \leq \frac{1}{n} \quad for \ all \ n \in \mathbb{N}. \qquad (2.2.39)$$

We have to prove that B is Reimann integral of ζ, given $\epsilon > 0$ suppose $y \in \mathbb{N}$ by using application of Archimedean property $y > \frac{2}{\epsilon}$. If $\dot{H}$ be tagged partition with $\| \dot{H} \| < \delta_y$ then

$$|S \left(\zeta; \dot{H} \right) - B|$$
$$\leq |S \left(\zeta; \dot{H} \right) - S \left(\zeta; \dot{G}_y \right) | + |S \left(\zeta; \dot{G}_y \right) - B|$$
$$\leq \tfrac{1}{y} + \tfrac{1}{y} < \epsilon, \qquad (2.2.40)$$

which shows that ζ be Reimann integrable with integral B.

$\square$

The following result gives the Reimann integrability of several calsses of functions:

Theorem 16. (*Squeeze Theorem*)
Suppose $\varsigma : [\alpha, \beta] \to \mathbb{R}$. Then ς be Reimann integrable $\Leftrightarrow$ for every $\epsilon > 0$, $\exists$ functions ϱ_ϵ and σ_ϵ be integrable function with

$$\varrho_\epsilon(u) \leq \varsigma(u) \leq \sigma_\epsilon \quad for\ all \ \ u \in [\alpha, \beta], \tag{2.2.41}$$

and such that

$$\int_\alpha^\beta (\sigma_\epsilon - \varrho_\epsilon) < \epsilon. \tag{2.2.42}$$

Proof. • ($\Rightarrow$) If we take $\varrho_\epsilon = \sigma_\epsilon = \varsigma$ for all $\epsilon > 0$. Which is required result.

• ($\Leftarrow$) Suppose $\epsilon > 0$, given that ϱ_ϵ and σ_ϵ be Reimann integrable, $\exists \delta_\epsilon > 0$ such that $\dot{R}$ be a tagged partition $\| \dot{R} \| < \delta_\epsilon$ then

$$\left| S\left(\varrho_\epsilon; \dot{R}\right) - \int_\alpha^\beta \varrho_\epsilon \right| < \epsilon \quad and \quad \left| S\left(\sigma_\epsilon; \dot{R}\right) - \int_\alpha^\beta \sigma_\epsilon \right| < \epsilon. \tag{2.2.43}$$

Then simple inequalities implies that

$$\int_\alpha^\beta \varrho_\epsilon - \epsilon < S\left(\varrho_\epsilon; \dot{R}\right) \quad and \quad S\left(\sigma_\epsilon; \dot{R}\right) < \int_\alpha^\beta \sigma_\epsilon + \epsilon. \tag{2.2.44}$$

By using (2.2.41), we get

$$S\left(\varrho_\epsilon; \dot{R}\right) \leq S\left(\varsigma; \dot{R}\right) \leq S\left(\sigma_\epsilon; \dot{R}\right), \tag{2.2.45}$$

whenever

$$\int_\alpha^\beta \varrho_\epsilon - \epsilon < S\left(\varsigma; \dot{R}\right) < \int_\alpha^\beta \sigma_\epsilon + \epsilon. \tag{2.2.46}$$

If $\dot{V}$ be another partition with $\| \dot{V} \| < \delta_\epsilon$, then

$$\int_\alpha^\beta \varrho_\epsilon - \epsilon < S\left(\varsigma; \dot{V}\right) < \int_\alpha^\beta \sigma_\epsilon + \epsilon. \tag{2.2.47}$$

Subtracting these above two inequalities and also using (2.2.42), we get

$$\left| S\left(\varsigma; \dot{R}\right) - S\left(\varsigma; \dot{V}\right) \right|$$
$$< \int_\alpha^\beta \sigma_\epsilon - \int_\alpha^\beta \varrho_\epsilon + 2\epsilon$$
$$= \int_\alpha^\beta \left(\sigma_\epsilon - \int_\alpha^\beta \varrho_\epsilon\right) + 2\epsilon < 3\epsilon. \tag{2.2.48}$$

Then by Cauchy Criterion implies that ς is Reimann integrable on $[\alpha, \beta]$.

□

2.3 Applications of Reimann Integrable

Now-a-days, Reimann integration is very important notion and very useful in other branches too. There are so many applications of Reimann integration and we present here just few of them.

Corollary 1. *Suppose ψ be integrable function on $[\alpha, \beta]$ and $m \leq \psi \leq n$, then*

$$m(\beta - \alpha) \leq \int_{\alpha}^{\beta} \psi \leq n(\beta - \alpha). \qquad (2.3.1)$$

Proof. Since ψ be integrable function and we know that Reimann integral over $[\alpha, \beta]$ on a constant function 1 is $\beta - \alpha$, therefore

$$m \leq \psi \leq n \qquad (2.3.2)$$

and then taking Reimann integral over $[\alpha, \beta]$ on these inequalities we get

$$\int_{\alpha}^{\beta} m \leq \int_{\alpha}^{\beta} \psi \leq \int_{\alpha}^{\beta} n \qquad (2.3.3)$$

which shows that

$$m(\beta - \alpha) \leq \int_{\alpha}^{\beta} \psi \leq n(\beta - \alpha). \qquad (2.3.4)$$

$\square$

Corollary 2. *Suppose ψ be integrable function on $[\alpha, \beta]$ then $|\psi|$ is integrable and*

$$\left| \int_{\alpha}^{\beta} \psi \right| \leq \int_{\alpha}^{\beta} |\psi|. \qquad (2.3.5)$$

Proof. Since ψ be integrable function on $[\alpha, \beta]$ means that for every $\epsilon > 0$, there exists a tagged partition $\dot{R}$ such that $\| \dot{R} \| < \delta_{\epsilon}$ then

$$S\left(\psi; \dot{R}\right) := \sum_{l=1}^{n} \psi(z_l) \left(u_l - u_{l-1}\right) \qquad (2.3.6)$$

exists. We know that for series, the following inequality is true

$$|S\left(\psi; \dot{R}\right)| := |\sum_{l=1}^{n} \psi(z_l)\left(u_l - u_{l-1}\right)| \leq \sum_{l=1}^{n} |\psi(z_l)| \left(u_l - u_{l-1}\right), \qquad (2.3.7)$$

by using this inequality we can prove easily the required result. $\square$

Corollary 3. *Suppose ψ be integrable function on $[\alpha, \beta]$ and that $\psi(u) = 0$ except for finite points $a_1, a_2, a_3, ..., a_n$ in $[\alpha, \beta]$ then $|\psi|$ is integrable and $\int_{\alpha}^{\beta} \psi = 0$.*

Corollary 4. *Suppose ψ be integrable function on $[\alpha, \beta]$ and also $\psi(u) = \varphi(u)$ except for finite points $a_1, a_2, a_3, ..., a_n$ in $[\alpha, \beta]$ then $|\varphi|$ is integrable and $\int_{\alpha}^{\beta} \psi = \int_{\alpha}^{\beta} \psi$.*

Corollary 5. *Consider ψ be function defined by $\psi(u) := u + 1$ for $u \in [0, 1]$ rational and $\psi(u) = 0$ for $u \in [0, 1]$ irrational, prove that ψ is not Reimann integrable.*

Corollary 6. *Consider that ψ be continuous on $[\alpha, \beta]$, that $\psi(u) \geq 0$ for all $u \in [\alpha, \beta]$ and that $\int_{\alpha}^{\beta} \psi = 0$. Show that $\psi(u) = 0$ for all $u \in [\alpha, \beta]$.*

Corollary 7. *If ϕ and φ be continuous on $[\alpha, \beta]$ and $\int_{\alpha}^{\beta} \phi = \int_{\alpha}^{\beta} \varphi$, show that there exists $a \in [\alpha, \beta]$ such that $\phi(a) = \varphi(a)$.*

Corollary 8. *Show that $\phi(u) := sin(\frac{1}{u})$ for $u \in (0, 1]$ and $\phi(0) := 0$ is Reimann integrable.*

Corollary 9. *If φ is a bounded and F is a finite set such that φ is continuous on $[\alpha, \beta]/F$, prove that φ is Reimann integrable.*

Corollary 10. *If ϕ is continuous on $[\alpha, \beta]$, $\alpha < \beta$ prove that there exists $a \in [\alpha, \beta]$ such that $\int_{\alpha}^{\beta} \phi = \phi(a)(\beta - \alpha)$.*

www.ingramcontent.com/pod-product-compliance
Lightning Source LLC
Chambersburg PA
CBHW041121180526
45172CB00001B/362